MIRROR OF HEAVEN

The sea: Serenity and passionate power

Charles Milton-Scott

A book in the series "Reflecting on the natural world"

Volume 3

Copyright © 2024 CHARLES MILTON-SCOTT

All rights reserved.

ISBN: 979-8339814610

PREFACE

To the series « *Reflecting on the natural world* »

Amidst the turmoil of modernity, a fundamental question arises: what can the contemplation of nature still bring to contemporary man? The answer is as simple as it is obvious: it is more relevant than ever. It is a beacon in the dark night of our time. Adrift in the maze of an ever-changing society, man has lost his bearings. The norms that once guided his thoughts and steps have disappeared in the complexities of modernity, leaving behind a sense of uncertainty and disorientation. Man is no longer at ease in this world of blurred contours. Values dissolve like mirages in the arid desert of postmodernity.

Above all, it is fear that insidiously creeps into the heart of man today. Fear of the present, fear of the uncertain future that looms on the horizon. This diffuse anxiety, fuelled by economic, political, environmental crises and social upheaval, is shaking the very foundations of our being. In this time of doubt and turmoil, man has also lost the faith that was once his guide and source of security and comfort.

In the midst of this internal and external turmoil, reflections on the natural world emerge as beacons in the stormy ocean of existence. They offer a refuge, a haven of peace, where man can regain his inner balance and reconnect with his roots. By contemplating the splendour of nature, by immersing himself in its unchanging beauty and millennia-old wisdom, man finds an echo of his fundamental aspirations and an answer to his existential questions.

The technologies that were supposed to contribute to his well-being and thus his happiness have led him into a spiral of confusion in which he finds no satisfaction and seeks an escape only in individual and social excess. The frantic pursuit of technological progress and an abundance of material goods has only deepened a gaping void in the human soul. It has left a sense of emptiness and disillusionment.

The first step in regaining emotional balance is to reconnect with the simplicity of nature. Reconnecting with its rhythms, feeling the earth beneath your feet, listening to the chirping of the birds, breathing the mountain air, admiring the dew on the flowers in the fields, are all means of soothing the troubled mind.

Beyond this personal quest for serenity, it is incumbent upon man to recognise his role as guardian of the biotope and therefore his responsibility to protect and nurture it. To help it flourish. Otherwise he is heading for his own demise. His emotional loss, then his existential loss and finally his disappearance. By neglecting the teachings of nature and exploiting its resources indiscriminately, man endangers his own survival and that of all life forms on the planet.

If man is to heal humanity and its earthly cradle, he must begin by healing himself. He must heal himself of his inner suffering, his fears and anxieties, in order to regain emotional and spiritual balance. He can do this by reconnecting with himself and listening to the teachings of nature.

Listening to the teachings of nature means learning to live in harmony with it. It means respecting its cycles and caring for its fragile ecosystems. It means understanding that every living thing, every plant, every animal, has its place and its role to play in the cosmic march. By observing how nature regenerates, adapts and always finds its balance, human beings can find answers to their own existential questions in order to heal not only their own wounded hearts and souls, but also those of humanity as a whole. This is the only way to preserve the legacy of life on Earth.

Charles Milton-Scott
September 2023

The sea is a place where every soul finds its own adventure.

William Shedd

Table of contents

1. Ulysses' odyssey: Calypso's cave 1
2. Untamed beauty 5
3. From calm to storm 9
4. Reflection and meditation 15
5. Seas and oceans of the world 19
6. Power and immensity 31
7. The source of life 43
8. Resilience and incompleteness: life lessons from an isolated pond 53
9. Fountain of riches 59
10. An inexhaustible source of inspiration.............. 67
11. The power of passion 81
12. The unattainable horizon............................. 87
13. The ocean in a drop of water 91
14. Perpetual motion 95
"Reflecting on the natural world" series 101
The author 105
Books by the author...................................... 107

x

1. Ulysses' odyssey: Calypso's cave

After years of wandering the Mediterranean, Ulysses, the brave Greek hero whose exploits are recounted in Homer's epic "The Odyssey", discovers the enchanting island of the nymph Calypso. It is an earthly paradise where the elements seem to embrace each other in eternal gentleness. Above all, the climate is so mild that it seems eternal. The sun shines with a golden light, bathing the island in a caressing warmth, while the sea, true to its nature, sparkles with an enchanting blue. Everything here seems to radiate beauty in its purest form at every moment.

Calypso, captivating and charming, holds Ulysses in her enchanting embrace. She promises him immortality and offers him the cup of eternal existence. Despite his burning desire to return to his homeland and the arms of his family, Ulysses is seduced by the sweetness of life on this island blessed by the gods. Each day unfolds in perfect harmony, the sea breeze carrying the sweet scent of flowers as a constant echo, while the birdsong fills the air with a soothing melody.

But Ulysses is a man of action, a fearless traveller whose destiny is tied to the sea. He knows that despite the comfort and beauty of Calypso's island, his true home is elsewhere, aboard his ship, amidst the majestic waves of the Mediterranean. The bright sun, sparkling sea and mild climate, however enchanting, cannot compete with the freedom and adventure that await him out there, nor with the vision of returning home.

So the god Hermes, messenger of the Olympian gods, is sent by Zeus to demand that Calypso release her hold on Ulysses. Although Calypso is reluctant to part with the man she loves, she understands that it is impossible to defy the will of the celestial powers. Eventually, she resigns herself to letting go and tells Ulysses that he is now free to continue his journey. She helps him build a raft, provides him with provisions for his journey, and then allows him to set sail across the open sea.

Freed from the enchantment that had held him captive, Ulysses resumes his journey to return to Ithaca, his home, with the promise of being reunited with his beloved Penelope.

Ulysses' odyssey: Calypso's cave

The story of Ulysses and Calypso highlights the eternal appeal of the sea, the sun and the mild climate. It reminds us that while nature offers earthly paradise, the call of adventure and travel remains rooted in the human soul. The sea embodies this unquenchable thirst for exploration, discovery and a return to our roots. Despite the beauty of calm skies and tranquil waters, an irresistible force drives the true journey to continue across the infinite sea of life.

≈≈≈

HEAVEN'S MIRROR

2. Untamed beauty

The sea is a force that never tires, never falls silent, never stops.

Antoine de Saint-Exupéry

The sea evokes a thirst for freedom, an irrepressible force that drives the spirit to cast off earthly moorings. The call of the open sea resounds like an enchanting melody, a symphony whose salty notes carry the soul towards distant horizons. It's an imperative call that awakens the sailor, urging him to set sail and explore the mysteries of the sea.

"Free man, you will always cherish the sea", wrote Charles Baudelaire[1]. These words echo from the depths of our soul. The sea, vast and mysterious, embodies freedom itself, a freedom that every human being longs to experience. It is the mirror of our desires, an endless expanse where the spirit can escape the constraints of the terrestrial world. It is fierce, wild and yet welcoming. It reminds us that true freedom lies

---

[1] In *Les fleurs du mal*, 1857.

not in the absence of constraints, but in the ability to lose oneself and be found again. To keep venturing towards new horizons, knowing that the unknown is always within reach.

A free man will always cherish the sea because it is the ultimate expression of what he seeks: a space where control and surrender coexist in harmony. It is in this duality that man finds his completeness, a way of understanding himself, of seeing himself as a being capable of both greatness and fragility.

Those who sail its waters know this well: every voyage is an adventure, where every wave can be an ally or an enemy. The sea is both a playground and a battlefield, a place where man faces his fears and his dreams. For those who appreciate it, it is a constant lesson in humility and courage. The distant horizons it unfolds before the eyes of dreamers are promises of discovery, new beginnings and infinite possibilities. This recognition of escape, of elsewhere, makes the sea a place where man can leave behind the invisible chains of society and his own doubts.

To cherish the sea is to cherish the very idea of freedom, adventure and infinity. It is to remember that, despite all the constraints of the modern world, there are still spaces where one can truly be oneself, where the soul can breathe and expand into the vastness. The sea calls us, reminds us of life, of the essential. That is why the free man will always cherish the sea.

Where the sky merges with the water, where the horizon seems to extend into eternity, man discovers a sense of infinity that feeds his insatiable desire to discover what lies beyond his dreams. The sea becomes a mirror of infinity, a backdrop against which the mind can project its wildest dreams.

Beneath the shimmering surface, the mysterious depths exert an irresistible pull on those who yearn to explore the mysteries of creation. Like an echo of buried desires, the sea beckons to immerse oneself in the unknown. It is a constant quest for understanding, an endless search for the treasures buried within ourselves.

The attraction of body and soul to the sea becomes a sacred communion with nature. The water caresses the skin, the wind ruffles the hair, each marine element creates an intimate bond between man and the ocean. In cherishing the sea, man discovers a timeless refuge where freedom finds endless resonance in the waves that cradle eternity.

The salty spray reveals the impetuous and captivating character of the ocean. The intoxicating scent of salt and iodine hangs in the air, permeating every breath with a marine essence that awakens the senses after a long slumber.

To the rhythmic pulse of a primordial beat, waves caress the rocks; ripples play on the cheerful notes of a melodic lapping that resonates like an enchanting lullaby. The regular ripple blends with the natural symphony of the depths, a perpetual pulse whose ephemeral surge leaves an indelible mark on the shore. It is an aquatic ballet, a universal dance between the earth and the sea, where the waves crash with hypnotic regularity before retreating like ballerinas backstage after a flawless performance.

The sea is not always calm and serene. Sometimes it transforms into a tumultuous scene where the elements rage. Deafening thunder crashes in the sky, accompanied by the primal roar of waves rising in a wild dance, while lightning splits the heavy clouds and streaks the ocean with a fleeting and terrifying light. A spectacle of indomitable force unfolds, a majestic display of nature's raw power. The elements rage in a frenzied trance, disregarding man's construction and ingenuity, sweeping away everything in their path.

From the salty spray to the wild choreography of the elements, every aspect of the sea's incessant morphogenesis tells a story. It is a universe in constant motion, where calm coexists with storm, where the untamed beauty of nature is revealed in every burst of foam and every breath of sea air. ≈≈≈

3. From calm to storm

O sea, how powerful you are in your anger and gentle in your calm!

Anonymous, Greek fragment

Since ancient times, countless stories and legends have celebrated the extraordinary power of the sea. One of these is that of Poseidon, the god of the seas and oceans in ancient Greek mythology.

Brother of Zeus and one of the twelve ruling gods of Mount Olympus, Poseidon ruled over the ocean depths. He had the power to calm or unleash the waves at will. One of the most famous legends depicts him building the walls of the majestic city of Troy, designed to be impregnable. Along with Apollo, Poseidon was forced by Zeus to serve King Laomedon of Troy. Together they built imposing walls around the city. Poseidon raised the walls with his divine strength, while Apollo brought herds of cattle to feed the city. However, Laomedon refused to pay Poseidon and Apollo for their services. Furious, Poseidon unleashed the fury of the seas against

Troy. He caused devastating floods and sent monstrous sea creatures to ravage the city.

The ancient Greeks used mythological legends to interpret natural phenomena and encourage respect for the untamable forces of nature. They saw the sea as an unpredictable and capricious force, capable of upsetting the destiny of mankind. The legend of Poseidon illustrates their admiration for the extraordinary power of the sea and how it could be equally invoked to inflict punishment or provide benevolent protection.

≈≈≈

The sea exerts an infinite fascination, a magnetic attraction. It jealously guards the secrets that have captured the heart of mankind since the dawn of time. A colossal force of nature, it evokes an impressive duality, a perpetual dance between raw power and exuberant vitality. It awakens a myriad of emotions and passions in the human soul. This mystical aura reveals itself in multiple facets that invite us to contemplate its grandeur, its role in the human journey and the thrilling emotions it evokes.

At first sight, the sea horizon stretches out like a canvas where sky and water meet. The waves whisper ancient stories, carry echoes of past civilisations, daring explorations and silent tragedies.

From calm to storm

It is a universe in constant motion, where the shifting hues of blue, green and turquoise captivate the senses and create living portraits of indescribable beauty. Above all, it symbolises the power of nature. Its boundless waters, its titanic waves, its ability to shape coastlines, generate storms and dance with the tides, all express an immense splendour and power. To gaze upon the sea is to confront one's own smallness in the face of titanic power. It is a mirror that reflects meditations on humility and modesty in the vastness of the universe.

The ocean is an inexhaustible source of life and fertility. It is home to billions of creatures, from the smallest to the most gigantic, and its biodiversity is a constant source of wonder. Marine ecosystems, the cradle of life, tell an eternal story of birth, regeneration and emergence. The ocean is a reminder of the continuous evolution and resilience of life. It also holds hidden secrets, its abyssal depths harbouring a mysterious world where strange creatures thrive in the darkness. Every dive into its waters is an adventure into the unknown, an exploration of magical territories.

The sea is an eternal muse for art, literature, philosophy and poetry. Artists have long found in it a metaphor for the human spirit, a reflection of the emotional spectrum that oscillates between serenity and exuberance. The ocean embodies the full range of human emotion, from contemplative reflection to fiery enthusiasm, from solemn respect to reverent fear and ecstatic admiration.

In spiritual and philosophical traditions it often symbolises the infinite, the unknown and the spiritual quest. It represents a mystery that drives the soul to seek knowledge, to dive into the depths of understanding and to marvel at the unfathomable depths of existence. It gives birth to countless myths and legends, fuels our perpetual quest for meaning and inspires us to meditate on our place in the cosmos.

The sea is a refuge with the power to guide our emotions. It evokes a sense of peace and serenity when its waters are calm, a thrill of excitement when they suddenly awaken, and panic when they flare up in Poseidon's rage. In these moments of unleashed passion, it embodies raw and untamed beauty, yet always returns to the calm and gentle tranquillity of waves caressing the shore. Its melodious song enchants our souls and transports us to distant dreams.

Watching the sun set over the ocean, feeling the sea breeze on your face or listening to the soothing sound of the waves lapping against the shore brings a comfort and peace that matches the infinite grandeur of the sea itself.

≈≈≈

From calm to storm

In its powerful mystery, the ocean reveals a true reflection of the depths of existence. Before us stretches a seemingly placid surface, an expanse of azure blue that sometimes turns turbulent when the wrath of the gods is brutally unleashed, hiding mysterious depths beneath. In the ocean's duality, oscillating between calm and turmoil, the complexity of the human condition is vividly portrayed.

The ocean offers an enchanting tranquillity. Its waves cradle the soul in an eternal dance, evoking peace and soothing serenity. Sunsets over the sea, with their soft golden hues, provide moments of contemplation and harmony that illuminate our inner quest. It is a sanctuary where the soul finds refuge, where the whirlwinds of daily life are calmed. In its quiet passion, it invites reflection and introspection.

The tides, controlled by celestial forces, express its changing moods and the unpredictability of nature. The ocean as a whole also embodies a certain lyricism in the stillness of its depths - a powerful sensuality that remains undisturbed.

In human life, moments of serenity and tranquillity are sometimes replaced by emotional storms and unexpected challenges. People experience periods of elation and indifference, face bright joys and dark sorrows. They navigate the waters of destiny whose uncertainty remains a mystery. The mixture of love, creativity and desire does not always manifest itself loudly, but these quiet passions are the sources that give life its meaning.

≈≈≈

4. Reflection and meditation

The sea is the embodiment of supernatural and marvellous existence.

Jules Verne

The vast expanse of the sea is an infinite mirror that reflects our inner journey towards serenity. Peaceful moments by the sea are open doors to our own primal inner universe.

The shores of the sea, with the soft sand beneath our feet and the salty scent of the sea spray caressing our skin, create a natural setting conducive to contemplation. In this place, where the crashing waves can shake the sky, the mind finds a space to calm and refocus. In these moments, the sea becomes the stage for our inner search for peace.

By the sea, meditation becomes an immersive experience. The soothing sound of moving waves blends with an inner music that invites quiet contemplation. The endless variations in the movement of the water reflect the richness of our thoughts.

These moments of tranquillity by the sea offer the chance to step away from the hustle and bustle of everyday life, to leave the worries of the outside world behind and explore the depths of the soul. Faced with the vastness of the ocean, the insignificance of our daily worries becomes apparent, prompting us to reflect on the ephemeral nature of life.

Like the layers of the ocean, the soul has many layers. Each holds its own mysteries and revelations. On the surface, the soul reflects tranquillity, much like the calm sea, whose smooth surface gives the illusion of absolute peace. However, like tides that ripple under the influence of invisible forces, human emotions create internal waves that are often invisible to the outside world. Past and present experiences form layers that tell a complex story engraved in the depths of the soul. Memories, whether sweet or bitter, are like corals that grow over time, creating a unique texture for each person, producing the essential oxygen for life.

The depths of the soul, like the unexplored depths of the ocean, hold mysteries. These depths symbolise the most intimate aspects of our being, from our most secret longings to our hidden fears. It is here, in these abysses, that the true treasures of the inner life lie, waiting to

Reflection and meditation

be discovered by those who dare to dive within themselves.

Complex and ever-changing, the currents are driven by the power of the winds that shape the surface. Relationships, experiences and chance encounters act like invisible winds that influence our path. Each interaction leaves a trace, each emotion triggers a wave that spreads across the ocean of our consciousness. These invisible forces create ripples in the calm waters of our mind, revealing hidden layers of our personality.

Like the tides, our emotions rise and fall with the pull of the moon, taking us through cycles of joy and sorrow, elation and despair. This constant ebb and flow reminds us that life is constantly evolving. To venture into the depths of the soul is to embrace this fluctuation, a dance between light and darkness where we learn to navigate the sometimes murky waters of our inner journey to find serenity and happiness.

The creatures of the deep sea, adapted to living in total darkness, symbolise our ability to find light even in the darkest moments. They teach us resilience, the ability to survive despite the most extreme conditions. We discover sources of courage and creativity that help us overcome challenges. By daring to face our fears and express our desires, we shed the masks we falsely present to those around us. We then come closer to our true essence, the truth hidden beneath the surface of appearances, the truth that connects us to the movement of the universe.

Just as ocean explorers gradually reveal the secrets of the deep, we are explorers of our own souls. We must brave inner storms and emerge enriched and transformed from the quest for self-knowledge. The mysteries of the ocean depths and those of the soul reflect each other in an infinite mirror of revelation.

Exploring the layers of the soul is an act of courage and curiosity, for each layer exposes a part of ourselves that we can choose to ignore or, conversely, accept, heal and celebrate.

≈≈≈

5. Seas and oceans of the world

The sea is a link between peoples, a path that no one can close.

Hesiod, Works and Days

Although the terms "sea" and "ocean" are often used interchangeably, they refer to distinct geographical entities. They also differ significantly in terms of size, geographical position, ecological characteristics, and human usage.

Oceans cover 71% of the Earth's surface and contain 97% of the planet's water. They are divided into five major bodies: the Atlantic Ocean, the Pacific Ocean, the Indian Ocean, the Arctic Ocean, and the Southern Ocean. With an average depth of around 3,700 metres, these expanses contain oceanic trenches, such as the Mariana Trench in the Pacific, which plunges to nearly 11,000 metres.

Oceans are continuous bodies of water that separate continents. They play a crucial role in regulating the global climate, the hydrological cycle, and global ocean currents, influencing weather patterns on a planetary scale. Seas, on the other hand, are located on the periphery of oceans or inland. They are often enclosed by continents or archipelagos, making them more accessible for human activities such as fishing, maritime trade, and tourism. However, their proximity to land exposes them more to the influence of neighbouring landmasses, affecting their salinity, temperature, and ecosystems. While seas vary in size and depth, they are generally shallower than oceans.

Oceans host ecosystems that range from surface waters rich in phytoplankton to abyssal depths where unique life forms thrive under extreme conditions. Ocean currents, such as the Gulf Stream in the North Atlantic, play a crucial role in nutrient distribution and climate regulation. Seas, being smaller and often more enclosed, have very specific ecosystems depending on their geographical location and degree of isolation. For instance, the Baltic Sea has low salinity due to freshwater input from rivers, while the Red Sea is known for its diverse coral reefs. Seas are also more susceptible to pollution and human activities due to their proximity to coastlines.

Oceans are vital for global trade, climate regulation, and as sources of food and natural resources. Their vastness and depth make them major subjects of scientific research, and much of their seabed remains unexplored. Seas, being more accessible, have played a crucial role in the history and development of human civilizations. They have served as essential trade routes and sources of sustenance through fishing. Seas often hold cultural and historical significance for coastal peoples, and they have inspired numerous legends, traditions, and practices.

There are several ways to classify seas, but the most common distinction is based on their location and connection to the oceans. Among the main categories are marginal seas, inland seas, and enclosed seas. Marginal seas, such as the North Sea or the East China Sea, are located on the periphery of oceans and are partially surrounded by landmasses. They are in direct contact with the ocean, allowing them to benefit from the hydrological exchanges that influence their salinity, temperature, and biodiversity. The North Sea, for example, is heavily influenced by the Atlantic Ocean, giving it characteristics similar to those of the ocean itself.

Inland seas, such as the Mediterranean or the Black Sea, are partially enclosed by land but remain connected to the oceans through straits or relatively narrow passages. These seas have unique ecosystems shaped by their relative isolation and the climatic and hydrological influences of their region. They have often played a crucial role in the history and trade of surrounding civilisations due to their accessibility and wealth of marine resources.

Finally, enclosed seas, such as the Caspian Sea or the Aral Sea, are completely surrounded by land and have no direct connection to the oceans. Due to their complete isolation, these seas have distinct characteristics, particularly in terms of salinity and ecosystems. They are often sensitive to environmental changes and the impact of human activities, which can lead to significant fluctuations in water levels and biodiversity. The Caspian Sea, for example, is the largest saltwater lake in the world, and its hydrographic characteristics are influenced by the rivers that flow into it and by intense evaporation due to the region's climate.

The Dead Sea is a remarkable example of an enclosed sea, surrounded by land and with no direct connection to the oceans. Located on the border of Jordan, Israel and the West Bank, it is famous for its extremely high salinity - one of the highest in the world - due to the intense evaporation in this arid region, combined with the lack of an outlet to another body of water. The high concentration of mineral salts makes the Dead Sea inhospitable to most aquatic life, hence its name. However, it is home to specialised micro-organisms that have adapted to survive in these extreme conditions. One notable example is the haloarchaea or halobacterium, a group of extremely halophilic archaea that thrive in high salinity environments. These microorganisms use light to produce energy in a process similar to photosynthesis, but adapted to high salinity conditions. Another example is the unicellular microalga Dunaliella salina, one of the few organisms able to survive in the Dead Sea. It resists salinity by producing large amounts of glycerol, a compound that helps to balance osmotic pressure; it is also known for its ability to produce carotenoids, which give it a reddish colour.

The water level of the Dead Sea has declined significantly in recent decades, mainly due to the over-abstraction of the rivers that feed it, especially the Jordan River, and accelerated evaporation due to the arid climate of the region. This drop in water levels is leading to the formation of depressions and the alteration of the unique ecosystem that surrounds this sea. The Dead Sea is also a site of historical and cultural importance and is renowned for its therapeutic properties. It attracts visitors for its mineral-rich mud baths and buoyant waters. Its status as a closed sea and its geographical location make it particularly sensitive to environmental changes and human activities, making it an important subject of study for scientists and ecologists.

≈≈≈

The polar oceans, located at the northern and southern extremes of our planet, represent another type of extreme environment where nature reveals both its power and its fragility. Surrounded by ice and subject to harsh climatic conditions, these seas play a crucial role in regulating the global climate and are home to unique and fragile ecosystems.

The Barents Sea, north of Norway and Russia, is one of the most famous polar seas. Despite its location beyond the Arctic Circle, it enjoys a relatively mild climate due to the influence of warm currents from the North Atlantic. This sea is an example of the resilience and adaptability of polar ecosystems. In summer, when the ice retreats, it becomes a vital breeding ground for many marine species such as cod, seals and various migratory birds.

In winter, the Barents Sea turns into an icy, cold desert where only the hardiest species can survive: Arctic cod, which have antifreeze proteins in their blood to prevent freezing in the frigid temperatures; Greenland and ringed seals, which dig dens in the ice where they can shelter and raise their young; polar bears, which roam the ice floes using their exceptional hearing and sense of smell to detect seals under the ice; and certain seabirds, such as the black-legged kittiwakes, which take advantage of areas where the sea remains partially ice-free due to Atlantic currents, allowing them to feed on fish.

The Barents Sea is of great geopolitical and economic importance. It is rich in natural resources, particularly oil and natural gas, which has led to significant economic interests and competition. In addition, its fish-rich waters make it an important fishing zone for neighbouring countries. However, this exploitation poses significant environmental challenges, as human activities can easily disrupt the delicate balance of the polar ecosystem.

Unlike the Barents Sea, the Ross Sea in the Southern Ocean is one of the most pristine and unexplored environments on Earth. Bordered by the vast ice expanse of the Ross Ice Shelf, the sea serves as a natural laboratory for scientists studying climate, ocean currents and polar biodiversity. The extreme conditions of the Ross Sea, with its freezing temperatures and permanent ice cover, create a unique habitat for species adapted to these challenges, such as the emperor penguin and the Weddell seal.

The Ross Sea is also a critical indicator of climate change. Phenomena such as glacial retreat and changes in ocean currents observed here have global implications for the planet's climate. The importance of protecting this unique ecosystem has led to the creation of the world's largest marine protected area, which covers much of the Ross Sea and aims to preserve its unique environment for future generations.

Located near the equator, tropical seas are havens of biodiversity and a paradise for researchers and tourists alike. They are characterised by warm waters, an incredible diversity of marine life and ecosystems that are as fragile as they are varied and spectacular. These seas play a crucial role in global climate, fisheries and tourism, and are sensitive indicators of environmental change. With their dazzling biodiversity and ecological, economic and cultural importance, tropical seas are ecosystems as precious as they are fragile.

The Coral Sea, north-east of Australia, is a prime example of the richness of tropical seas. It is home to the Great Barrier Reef, the world's largest coral reef, visible from space. The reef is a complex and diverse ecosystem, home to thousands of species of coral, fish, molluscs and other marine life. The warm, clear waters of the Coral Sea provide ideal conditions for their growth. Coral reefs are often referred to as the 'rainforests of the sea' because of their enormous biodiversity. But they are also extremely sensitive to environmental change. Ocean warming, acidification and pollution are major threats to their survival. Coral bleaching, caused by thermal stress, is an alarming sign of the fragility of this ecosystem.

Seas and oceans of the world

The Arabian Sea, bordering the Arabian Peninsula and India, is another tropical sea with unique characteristics. It is characterised by high temperatures and significant evaporation, which affect its salinity and currents. It also serves as an important maritime route linking the economies of Asia and the Middle East with those of Africa and Europe. Ocean currents, such as the Oman Current, play a key role in the distribution of nutrients and support a thriving fishing industry. Its shores are lined with mangroves and coral reefs, which provide essential habitats for marine life and protect the coastline from erosion. However, like many tropical seas, it is vulnerable to the effects of pollution, including oil spills and plastic waste.

Tropical seas are also centres of intense human activity. They attract millions of tourists each year who travel to enjoy sandy beaches, crystal clear waters, coral reefs and marine life. Fishing in these seas is vital to the food security of many communities. Both artisanal and commercial fishing rely on the rich fish stocks in these waters, but overfishing and destructive practices threaten their sustainability. Tourism is a vital source of income for many coastal economies, but it must also be managed sustainably to avoid damaging the environment.

While polar seas, with their freezing temperatures, have lower salinity levels due to the influx of freshwater from melting ice, tropical seas, influenced by high evaporation rates, tend to be saltier. This variation in salinity affects water density, ocean currents and nutrient distribution, and has a significant impact on local biodiversity. Although less diverse than tropical seas, polar seas are vital for iconic species and unique ecological processes such as whale migrations and sea ice formation.

The comparison between polar and tropical seas highlights the diversity of marine environments on our planet. However, each sea, with its unique challenges and ecological assets, plays a critical role in the overall health of the oceans and the diversity of marine life.

The management and conservation of seas requires approaches tailored to their specific characteristics. Conservation efforts in polar seas often focus on protecting against the effects of climate change and conserving critical habitats. In tropical seas, sustainable resource management, coral reef protection and tourism regulation are priorities to ensure the health and resilience of ecosystems.

It seems increasingly clear that humanity must take responsibility for the damage it is doing to the planet, but what should be a simple truth is beginning to feel like an unattainable utopia.

Seas and oceans of the world

Despite the sporadic efforts of environmental organisations and a few committed individuals, the collective will to address the ecological crisis remains elusive.

While awareness of the problem has grown, meaningful action is often undermined by apathy, short-term interests and political inertia. The moral responsibility to protect and restore the Earth should be self-evident, yet the gap between awareness and action continues to widen.

As we witness unprecedented environmental degradation - forests burned, oceans polluted, species driven to extinction - our duty as stewards of the planet feels ever more urgent. Yet the fragmented efforts of a few passionate advocates are not enough. Without a global shift in consciousness and a unified resolve to act, the idea of repairing the damage we have done is drifting further into the realm of idealism.

The question remains: will humanity rise to the challenge, or will our inaction determine the legacy we leave behind?

≈≈≈

HEAVEN'S MIRROR

6. Power and immensity

The sea is as deep in calm as it is in storm.

John Donne

During storms, the waves can sometimes be over thirty metres high. The gusts of wind, the songs of the tides and the movements of the earth create titanic forces. The sea, both an enchanting spectacle and a formidable source of danger, has countless ways of revealing its power.

Ocean storms, tsunamis and hurricanes are all examples of these devastating forces. Such natural phenomena dramatically affect coastal communities, causing flooding, destruction and human tragedy. Despite its inspiration, the sea reminds us to remain humble in the face of nature's vagaries.

A natural force beyond human imagination, the ocean's waters reveal both the fragility of our existence and the strength of our connection to the earth.

≈≈≈

Over the ages, the waves and tides have been relentless craftsmen, shaping the coastline. When the power of the sea caresses the rocks, it acts like a giant sculptor. The water, carrying fragments of sand and rock, patiently nibbles away at the geological contours, transforming the coastal cliffs into blocks subject to constant erosion. Under the chisels of this natural force, once impenetrable cliffs give way to golden sandy beaches or intricately carved rock formations. The white chalk cliffs of Étretat or Dover are majestic testimony to this constant work of erosion, where the sea occasionally unleashes its power in spectacular displays of strength.

As the sea erodes the coastline, it adds new features. The waves form natural arches, carve secret tunnels and erect rock pillars. The sea stacks on the Isle of Skye in Scotland are living creations, the result of millions of years of sea and stone working together.

Both creator and destroyer, the sea embodies the enduring influence of time on the landscape. Its works, visible around the world, are a testament to the power and beauty of nature in constant evolution. Although slow and often imperceptible on a human timescale, these changes remind us that the natural world is in constant motion, subject to forces beyond our comprehension. The coastlines we see today are merely stages in an ongoing story — a story in which the sea tirelessly redraws the face of the earth, creating works of art that are both fleeting and eternal.

≈≈≈

A capricious muse, the vastness of the blue inspires adventurous souls to great feats while reminding them of the fragility of existence. Those who manage to reconcile this paradox draw on an unyielding strength that echoes the primal elements.

It whispers tales of lost sailors to those who wander its shores. Writers, painters, poets and musicians find in its shadows an inexhaustible source of inspiration. Sunsets brush its horizon, bathing the world in a golden glow - a celestial canvas of eternal beauty tinged with truth. Its waters stretch endlessly, an ocean of possibilities, a mirror of dreams and contemplation, where meditation finds the serenity needed to attain inner peace. Its tranquil waters cradle our souls.

Beneath its azure surface lies immeasurable power. When the sea turns into a merciless fury, roaring with titanic force, it becomes the guardian of destruction, a colossal creature that devours all who dare challenge its power with cold, unforgiving arms.

The sea is the salty scent of spray on the skin, the softness of sand underfoot and the crashing of waves shaking the heavens. The apparent duality of calm and storm, inspiration and destruction, embodies the complexity of human existence, where emotions ebb and flow like the tides. Sometimes calm and soothing, sometimes devastating. In the depths of humanity, it is never certain who will emerge victorious. Yet the sea always returns to the calm in which the stars are reflected.

This duality does not mean conflict; it is the very essence of the breath of life. Like the ever-changing waves, life is made up of moments of calm and moments of turmoil. The depths of the sea have long reconciled this duality. They show us the way to find peace in the heart of the storm and to recognise the beauty in every facet of existence. The sea is both our inspiration and our mirror, and the union of opposites reveals a truth about the ephemeral nature of the universe.

≈≈≈

From time immemorial, countless events, stories and legends have shown how the immense power of the seas has shaped destinies, determined victories and defeats, and highlighted human vulnerability in the face of the majesty and brutality of the natural elements.

One such legend is that of Atlantis, which tells of a continent submerged by the ocean in the wrath of the gods. According to the story, Atlantis was a powerful and advanced

Power and immensity

civilisation located beyond the Pillars of Hercules [2]. The Atlanteans were considered exceptional beings, endowed with extraordinary knowledge and skills. However, their wealth and arrogance blinded them and led them to challenge the gods.

Legend has it that Zeus, the king of the gods, and Poseidon, the god of the seas and oceans, were angered by the Atlanteans' attitude. In response to their insolence, Zeus decided to punish this civilisation. He called upon Poseidon, who unleashed the fury of the oceans to punish the Atlanteans.

The sea, once a friend of Atlantis, turned against it. Terrible earthquakes, floods and devastating tsunamis were unleashed. Huge waves engulfed the lands, submerging its cities, palaces and flourishing cultures under relentless masses of water.

Atlantis was quickly engulfed by the ocean. It disappeared into its depths after an apocalyptic cataclysm. What had once been an advanced and prosperous civilisation became nothing more than a myth passed down through the generations.

For all their progress and arrogance, humans remain powerless against the forces of nature and the elements. The sea can be both

---

[2] A place that is generally associated with the Strait of Gibraltar.

benevolent and merciless. Its immense power gives life and takes it away in devastating ways. It teaches humility.

Tsunamis have had a significant impact on the history of the Aegean region. For example, the volcanic eruption of Thera (Santorini) around 1600 BC probably triggered a massive tsunami that affected the Minoan and Mycenaean civilisations. Such natural disasters have left an indelible mark on the collective imagination and may have inspired legends such as that of Atlantis.

In December 2004, a tsunami of astonishing proportions struck the coastal regions of the Indian Ocean. A powerful undersea earthquake had shaken the seabed with overwhelming force. The aftermath of this natural disaster was a cruel reminder of the ocean's ability to generate deadly waves.

The earthquake, which registered between 9.1 and 9.3 on the Richter scale, occurred off the northwest coast of Sumatra, Indonesia. The converging tectonic plates in this region released a colossal amount of energy, causing a sudden deformation of the ocean floor. This deformation caused the seafloor to shift vertically, pushing a massive volume of water towards the ocean surface. This sudden surge of water created huge waves that travelled across the Indian Ocean at an alarming rate.

Power and immensity

As these waves reached the coastlines of various countries, they turned into walls of water, submerging everything in their path. Peaceful coastal communities were engulfed in a matter of moments. As the sea violently retreated from the land, it took everything in its path and left behind a devastated landscape. The human losses were immense, and the material damage incalculable.

While this event was a brutal reminder of the chaotic power of the sea, it also triggered a massive international response in terms of relief efforts, disaster preparedness and scientific research. The lessons learned have led to a better understanding of the risks associated with undersea earthquakes and the establishment of early warning systems. Even in the midst of tragedy, the sea inspires humanity.

A historical episode known as the *Kamikaze* (meaning "divine wind" in Japanese) occurred during an attempted invasion of Japan by Mongol forces led by Emperor Kublai Khan. He launched two campaigns to invade Japan: the first in 1274 and the second in 1281. The first invasion was a partial failure, but the second is the most famous for its dramatic outcome.

During this second attempt, the Mongol fleet ready to invade Japan consisted of thousands of ships. As the Mongol forces prepared to land, a violent storm hit the sea, destroying much of the fleet. The storm caused great damage and marked a decisive turning point in the war. The Mongols were forced to retreat, ending their invasion attempts. The Japanese attributed the storm to the intervention of the kamikaze, divine winds sent by the gods to protect their country. This event strengthened the belief in Japan that the nation was untouchable because it was protected by the gods.

The story of the kamikaze became a legend, leading to the later use of the term "kamikaze" to describe the Japanese pilots who sacrificed themselves by attacking Allied ships during World War II. These pilots, driven by an extreme sense of duty and sacrifice, saw themselves as cherry blossoms falling from the sky. The cherry blossom, a fleeting yet powerful symbol of the beauty and fragility of life, embodied the purity of their mission and their acceptance of fate. In launching these desperate attacks, they honoured the spirit of their ancestors by protecting their homeland and fulfilling their ultimate duty with the grace and transience of petals carried by the wind. This poetic and tragic vision still resonates in the collective memory of Japan, where the sacrifice of these young men is often surrounded by an aura of respect and melancholy.

The powerful symbolism that emerges intertwines elements of the sea, air and sacred land. In their final flight, these pilots merged their essence with the elemental forces of the world. Their fleeting and fatal passage reflected the eternal cycle of life and death inscribed in the skies and seas surrounding Japan's islands. The beauty of their gesture, tragic as it was, lay in this union with nature, an ultimate sacrifice to the universe, where each kamikaze flight became a silent prayer for lost harmony, with the ocean waters serving as an eternal tomb.

≈≈≈

A capricious goddess, the sea has often revealed its dual nature. Nourishing and sustaining mankind, it sometimes snatches them away without warning, dragging them into its depths. On that day in 1964, as the sun set peacefully on the horizon, no one could have foreseen the events that were about to unfold.

A military vessel, like so many others, had anchored for the night not far from a Normandy port. The anchorage had been calm, almost too calm, like a deceptive prelude to the storm to come. After two weeks at sea, some of the crew had been given shore leave to relax. The sailors were delighted at the prospect of setting foot on solid ground again, eager to forget the harshness of life at sea, if only for a short while. A young and enthusiastic midshipman confidently led the eleven men into the launch. There was laughter and banter as they imagined the pleasures of their first sips of wine in a welcoming tavern in the small coastal town.

But the sea, cunning as ever, had other plans. When the launch was halfway to the harbour, a violent wind suddenly arose, as if blown by an invisible force. In an instant, the sky, previously clear and serene, was covered with dark, heavy clouds. The waves, once calm, rose into towering mountains of water, threatening to engulf everything in their path.

The crew on deck watched helplessly as the tragedy unfolded. The small launch, a fragile vessel against the fury of the elements, struggled desperately. But the waves were relentless, and with a terrible crash they overtook it. The launch capsized, throwing the sailors into the raging waters.

The Bosun, an old man whose skin was weathered by years at sea, and the usually stoic Captain sprang feverishly into action, ordering the ship's engines to start and the rescue operation to begin. The searchlights were switched on, piercing the night with their pale beams, searching for any trace, any sign of life. The ship circled the stormy sea like a desperate giant, attempting to find its lost crew.

But that night, the sea refused to give back what it had taken. All night long the men searched the waves, calling out in vain for their comrades, their voices drowned by the howling wind. Nothing. Not an echo, not a cry. The sea

Power and immensity

had swallowed the sailors, along with any hope of seeing them alive again.

When the faint grey dawn broke, it revealed a desolate shore littered with debris. The remains of the launch lay scattered, silent witnesses to a tragedy. The bosun, who had seen and experienced so much, could not hold back his tears. And the captain, that unshakable rock, wept too, shedding the bitter tears of those who carry the weight of lost lives on their shoulders.

That day the sea had taken, for no apparent reason, no justification. It had reminded everyone of its fierce nature, beyond human understanding. In the silence of dawn, where only the waves seemed to murmur, one certainty remained: the sea, both beautiful and cruel, makes no distinction. It gives, it takes and sometimes it takes everything.

≈≈≈

HEAVEN'S MIRROR

7. The source of life

The sea is the mother of all forces and all forms of life.
Jules Michelet

In its mystery and vastness, the sea is the primordial cradle of all life on Earth. It is both womb and nurturer, a realm of depths where existence finds its roots. To gaze upon its waves is to feel an ancestral connection, a memory of the first pulses of life in its ancient waters.

Billions of years ago, in this aquatic womb, the first cells stirred, fragile and tiny. They began their dance of survival and multiplication, carried by the ocean currents, setting in motion the biochemical reactions that would lead to the development of life. The sea, with its minerals and nutrients, provided the perfect environment for this genesis. Every wave, every tide, whispered the secrets of life, inviting these primitive forms to grow and evolve.

Coral reefs, realms of biodiversity, are living testimony to this primordial abundance. Here, the sea celebrates life in an explosion of colours and shapes, where millions of creatures live in harmony. Fish, crustaceans, sea anemones, algae and countless other creatures perform a never-ending ballet, a hymn to fertility and the continuity of life. Every crevice, every fissure holds secrets, stories of survival and symbiosis inscribed in the living stone of coral.

The sea is not only the cradle of life on earth, but also a generous provider. The vast oceanic spaces are home to countless marine species, from the tiniest plankton to the majestic whales. Invisible to the naked eye, plankton form the basis of a complex food chain. It converts sunlight into energy, sustaining an entire ecosystem. Graceful giants, whales filter tonnes of these tiny creatures through their baleen, linking the tiny to the immense in a perfectly orchestrated ballet of interdependence.

The unchanging rhythm of the tides governs life along the coasts. The constant movement of the waves brings in and removes nutrients, regenerating estuaries and mangroves — critical ecosystems where fresh water meets salt water. Migratory birds, fish and crustaceans depend on these rich, dynamic areas for food and reproduction. Every ebb and flow is a promise of renewal, a vital breath that animates these ever-changing landscapes.

Ancient sailors, scanning the infinite horizon, saw in the waves a reflection of the mysteries of creation. Their stories tell of sea gods and

The source of life

fantastic creatures, symbols of the power and the unknown of this fundamental element. In its immensity, the sea embodies the mystery of origin, the alpha of life's great adventure.

Today, as we explore its depths with advanced technologies, we continue to discover unexpected forms of life and hidden worlds in the abyss where light cannot reach. Each discovery reminds us that the sea is an inexhaustible source of life and mystery.

≈≈≈

Several theories have been proposed to explain the origin of life on Earth, each approaching the question from a different angle based on different scientific observations. The most widely accepted theory is that cyanobacteria played a crucial role in the development of Earth's oxygenated atmosphere, which allowed more complex life forms to evolve. Cyanobacteria, once known as "blue-green algae", are among the oldest known photosynthetic organisms. They emerged about 3.5 billion years ago and have the unique ability to perform oxygenic photosynthesis, a process that uses solar energy to convert carbon dioxide and water into glucose and oxygen.

Originally, cyanobacteria thrived mainly in aquatic environments, especially oceans and lakes. They colonised shallow waters where sunlight was sufficient to support the photosynthetic process. These microorganisms formed microbial mats on sea beds, lagoons and intertidal zones.

Their ability to photosynthesise not only facilitated their development and survival, but also played a crucial role in transforming the Earth's atmosphere. By releasing oxygen, they set the stage for the evolution of increasingly complex life forms.

Before cyanobacteria appeared, the Earth's atmosphere was rich in carbon dioxide and poor in oxygen. Around 2.4 billion years ago, the accumulation of oxygen produced by cyanobacteria led to an event known as the Great Oxidation Event. This event radically changed the Earth's atmosphere, allowing for aerobic respiration — a more efficient process of energy production that uses oxygen.

The increase in atmospheric oxygen encouraged the development of more complex organisms and played a crucial role in the evolution of life on Earth. This abundance of oxygen allowed eukaryotes — including plants, animals and fungi — to emerge and evolve. Aerobic respiration, made possible by the presence of oxygen, provided a more efficient source of energy, supporting increasingly complex cellular and multicellular structures.

In addition, the oxygen released by cyanobacteria had a major impact on the planet by contributing to the formation of the ozone

The source of life

layer[3] in the upper atmosphere. This ozone layer has a vital function: it protects the Earth's surface from the sun's harmful ultraviolet rays. This protection allowed different forms of life to colonise the land, paving the way for the biodiversity we know today.

Stromatolites, fossilised structures formed by the activity of cyanobacteria in these primitive aquatic environments, are some of the oldest fossil evidence of life on Earth. They show that cyanobacteria thrived in warm, mineral-rich waters, often in symbiosis with other microorganisms. In addition, isotopic analyses of ancient rocks reveal chemical signatures consistent with an oxygen-enriched atmosphere at the time.

The cyanobacteria origin of life theory explains how these microscopic organisms not only survived in primitive conditions, but also created an environment conducive to the emergence of complex life forms as we know them today.

Over the decades, however, other scientific theories have emerged. The first, the "primordial soup" theory, was proposed by Alexander Oparin and John Haldane in the 1920s. It suggests that life began in a "soup" of simple organic compounds formed from gases in the primitive atmosphere (methane, ammonia, hydrogen and

---

[3] Ozone is a molecule composed of three oxygen atoms (O_3).

water vapour) under the influence of energy sources such as lightning or ultraviolet rays. The Miller-Urey experiments showed that amino acids, the building blocks of proteins, could form under these conditions.

In 1952, Stanley Miller and Harold Urey conducted an experiment at the University of Chicago. Their work aimed to simulate the supposed conditions of the Earth's primitive atmosphere in order to understand how the first organic molecules, the precursors of life, might have formed. They mixed gases thought to represent Earth's early atmosphere (mainly water vapour, methane, ammonia and hydrogen) in a sealed apparatus. They then passed electrical discharges through the mixture to simulate lightning. After a week, they discovered that amino acids, the building blocks of proteins, had spontaneously formed in the mixture.

This experiment demonstrated for the first time that organic molecules essential for life could form from simple chemical compounds in conditions similar to those on the early Earth. Their work significantly influenced theories of the origin of life and remains iconic in studies of prebiotic chemistry.

Another theory, the hydrothermal vent theory, suggests that life may have originated around underwater hydrothermal vents, where

The source of life

favourable conditions such as heat, minerals and chemical compounds could have enabled the formation of complex organic molecules. Hydrothermal vents provide a source of chemical energy that could have powered the earliest living organisms.

The panspermia theory proposes that life on Earth could have an extraterrestrial origin. Microorganisms or organic molecules necessary for life could have been delivered by meteorites, comets or cosmic dust. However, this theory does not solve the question of the origin of life; it simply shifts it elsewhere in the universe.

The mineral surface theory, proposed by Graham Cairns-Smith in the 1960's, suggests that the first organic molecules may have accumulated on mineral surfaces, such as clay. These minerals may have catalysed the chemical reactions necessary to form complex organic molecules by acting as a matrix for the initial prebiotic reactions.

The primordial RNA theory, proposed by Nobel Prize winner Walter Gilbert in 1986, suggests that the earliest forms of life were based on RNA[4], a molecule capable of both storing genetic information and catalysing chemical reactions.

---

[4] RNA (ribonucleic acid) is an essential molecule in the cell that plays a crucial role in the transcription and translation of genetic information: it directs the synthesis of proteins from DNA.

RNA may have played a crucial role before the evolution of DNA and proteins. Experiments have shown that RNA molecules can replicate and evolve, supporting the idea that RNA could be the ancestor of modern biological molecules.

Lipid and membrane theory proposes that the formation of lipid membranes, structures similar to cell membranes, was a critical step towards life. These membranes would have allowed the creation of isolated compartments in which chemical reactions could occur more efficiently, facilitating the emergence of complex biochemical systems. The geochemical cycles theory, proposed by Michael Russell and others in the early 2000's, suggests that the geochemical cycles of elements such as carbon, nitrogen and sulphur, interacting with the Earth's primitive environment, favoured the formation of organic molecules and prebiotic structures. Temperature and chemical concentration gradients would have played a crucial role in the emergence of the first living systems.

Recently, scientists discovered a surprising "black oxygen" from metallic pebbles more than 4 kilometres deep in the Pacific Ocean. This observation could challenge current theories about the origins of life on Earth. The discovery was made in the abyssal plain of the Clarion-Clipperton Rift Zone, a vast region of the Pacific Ocean floor. It lies between the Clarion and

The source of life

Clipperton Islands, south of the Hawaiian Islands, and extends for several thousand kilometres from the south-east of the Baja California peninsula in Mexico to almost the equator. This zone is particularly known for its abundance of polymetallic nodules, which are deposits rich in precious metals such as manganese, nickel, copper and cobalt. These nodules are scattered across the seabed and are of great interest to the marine mining industry, although their exploitation raises environmental concerns.

In samples taken by the Scottish Association for Marine Science (SAMS), researchers observed an unexpected increase in oxygen levels in the seawater above the sediments, even in the absence of light and photosynthesis. The nodules, which can be likened to "batteries in rock", could produce oxygen by electrolysis, using an electric current to split water into hydrogen and oxygen. The discovery is prompting a rethink of how oxygen is formed and, by extension, the origin of life on Earth. Researchers suggest that life may have begun elsewhere than on dry land, and that similar processes may exist on other 'ocean worlds' such as the moons Enceladus and Europa, satellites of Saturn and Jupiter respectively, which may be conducive to extraterrestrial life. These findings also aim to better regulate deep-sea mining.

Each of the existing theories proposes a different mechanism to explain how simple molecules could have evolved into complex biochemical systems capable of reproduction and metabolism, laying the foundation for life as we know it.

None of these theories necessarily excludes the others, and it is possible that several processes combined to contribute to the emergence of life on our planet.

≈≈≈

The sea, with its eternal waves and currents, continues to nourish and inspire awe. It reminds humanity that we all come from its primordial waters. The sea holds the memory of the Earth, the key to our origins, and the promise of a future where life in all its forms will continue to flourish.

≈≈≈

8. Resilience and incompleteness: life lessons from an isolated pond

Resilience is not a condition but an ability; it is dynamic and built through trial and error.

Boris Cyrulnik

Sometimes, when a large wave crashes over a rocky shore and the tide quickly recedes, a small pool is left isolated from the ocean. This phenomenon creates a temporary marine microcosm, an oasis of aquatic life that becomes a theatre of resilience. Tiny fish swim frantically to escape and then adapt to their new, confined environment. Small crabs explore every nook and cranny in search of food, while seaweed, now deprived of the constant movement of the waves, sway gently in the still water. Every creature in this pool must endure the heat of the sun, which warms and evaporates the water, gradually shrinking its habitat. Life persists and continues against all odds.

The incomplete biotope of the tide pool, a truncated slice of life, becomes a poignant stage for the struggle for survival. Isolated from the vast ocean, the pool becomes a world unto itself, a micro-reality where each organism must do whatever it can to continue its existence. The intense heat of the sun threatens to dry out the water, making the habitat increasingly hostile. Yet life perseveres, adapting, fighting and finding ingenious ways to survive until the next tide comes to free them from their temporary captivity — if they manage to survive until then.

This phenomenon reminds us of the fragility and tenacity of life. Even under adverse conditions, nature finds ways to persevere. The isolated pools serve as natural laboratories where we can observe the mechanisms of adaptation and survival. They symbolise the ingenuity of life, its ability to resist even in the most precarious circumstances.

In a broader sense, these marine microcosms are metaphors for the human condition. Like the organisms in the tank, humans can find themselves isolated and faced with unforeseen challenges. Yet it is often in these moments of crisis that we discover our ability to adapt, to innovate and to persevere. The small tidal pools are a visual reminder of life's resilience, of perseverance in the face of adversity, and of the

Resilience and incompleteness: life lessons from an isolated pond

beauty that can emerge from even the most hostile environments.

The small pool created by the random action of natural elements, isolated and vulnerable, is a scene where much of the story of life itself is played out. A microcosm where every element, every creature struggles to exist, while teaching us lessons in courage in the face of adversity.

So it is with human existence, often incomplete and marked by periods of isolation and struggle. Like the pool created by the receding tide, our lives are often punctuated by moments when we feel cut off from our usual environment, faced with unforeseen challenges and forced to adapt to new realities. These periods can occur during personal crises, losses or major changes that leave us disoriented and vulnerable.

The incompleteness of human existence manifests itself in unfinished dreams, unfulfilled aspirations and interrupted relationships. We are often faced with situations in which we must struggle against forces, both external and internal, to maintain a semblance of normality and progress. The harsh reality, like the sun for the small pool, sometimes seems too intense and threatens to evaporate our hopes and resources.

It is in these moments of incompleteness and isolation that our true strength is revealed. The adaptability of the human spirit, its ability to find ingenious ways to survive and even thrive under adverse conditions, is one of the most remarkable features of existence. We learn to persevere in the troubled waters of life, to find pockets of support and hidden resources within ourselves and our environment.

Like the organisms in the isolated pool, we often have to reassess our priorities, adjust our expectations and find ways to carry on despite the constraints. This constant struggle to adapt and survive forms our character, shapes our personality and gives us the strength to carry on. Every challenge we overcome, every adversity we face, contributes to our personal growth. They give us a better understanding of ourselves and of our environment.

The imperfection of the human condition is not a weakness, but a condition that drives us to strive, to surpass ourselves and to thrive in imperfection. It reminds us that perfection is not an attainable state, but an ideal towards which we strive, and that it is in seeking and striving that our true nature finds its full measure.

The situation is particularly dramatic when a child grows up in an environment that is not conducive to his or her development, where the conditions do not allow him or her to fully develop emotionally, physically or intellectually. This environment may be characterised by a lack of affection, physical or psychological violence, extreme poverty or limited access to education and health care. The effects of these

Resilience and incompleteness: life lessons from an isolated pond

deficiencies are devastating when they impede a child's normal development and leave scars that persist into adulthood.

In such a context, the child is often forced to adapt to realities that are far too harsh for his or her age. The lack of emotional support and security creates feelings of insecurity, fear and low self-esteem. These young people are more likely to develop attachment disorders, academic difficulties and risky behaviour. Worst of all, these circumstances stifle their potential, depriving them of opportunities for personal development and a better future.

Far from being simply a matter of physical survival, the development of a child in an inadequate environment touches the very heart of what it means to be human: the ability of every human being to grow with dignity, to fulfil his or her potential and to contribute to society.

Just as the small pool isolated by the tide is a microcosm of relentless life, our own lives, with their ups and downs, their moments of isolation and struggle, are a testament to human perseverance. It is this perseverance that allows us to transform our sense of incompleteness into a journey of discovery and growth.

Unfortunately, many cases are far from successful, and the reality of human perseverance is often marked by trials that seem insurmountable. For some, the obstacles are too many or too heavy to bear, and the struggle to find meaning or direction can feel like an endless struggle against the current. In these situations, when hope falters, the shadows of doubt, fear or disappointment obscure the path to personal discovery.

Even in these moments of despair, perseverance is not always measured by spectacular victories or transformations. Sometimes it is simply the ability to carry on, to get back up after a fall, or to find the strength to face the next day. It is in these small actions, these quiet acts of courage, that the essence of human resilience is found.

Just as the isolated pool depends on the tides and surrounding elements, we are shaped by our environment, our relationships and the support we receive. For each person's story to have a chance of becoming a success story, the rising tide of a support network must overwhelm the inhospitable rocks. It brings the love and encouragement of loved ones, access to resources for personal development, and a community that values mutual aid and compassion.

Even if not everyone reaches the desired shores and peaks, perseverance continues to play a vital role in each person's story. It drives us to seek ways to grow through our experiences, whether marked by success or adversity, and to transform our journey into a search for meaning and human connection. ≈≈≈

9. Fountain of riches

The ocean is the last wild place on earth, an unexplored reservoir of biodiversity and wealth.

Sylvia Earle

The seas and oceans are a treasure trove in many ways. First, they play a crucial role in regulating the myriad climates that affect the biosphere. They absorb the sun's heat and distribute it over the planet's surface. In their vast thermal reservoir, they maintain balanced temperatures. They regulate rainfall, dictate the paths of ocean currents and shape the diversity of weather patterns. By trapping vast amounts of carbon dioxide, they form a barrier against the onslaught of climate change. This regulation of atmospheric and hydrothermal variability helps to maintain environmental conditions at levels compatible with life on Earth.

The ocean breathes life. Its mysterious depths are home to an extraordinary variety of wildlife. From microscopic phytoplankton to majestic whales, from seaweed to coral, from fish to dolphins, from molluscs to crustaceans, the sea is teeming with life. This biological richness is the beating heart of the oceans.

In the clear waters of the tropics, coral reefs provide shelter and food for countless marine species. From colourful fish to majestic sharks, from slow-moving turtles to delicate corals, from busy crustaceans to seaweed swaying with the currents, life flourishes. Sadly, these magnificent reefs are under threat, threatened by climate change, pollution and relentless overfishing.

Coastal ecosystems are full of mystery. Salt marshes, mangroves and estuaries act as nurseries for marine life. These natural filters maintain coastal water quality by trapping sediments and excess nutrients. By storing carbon dioxide, they help to mitigate the effects of climate change.

The pelagic zone, those vast, uninterrupted expanses of open ocean, remains a mystery. Plankton, the fragile backbone of the food chain, drift with the whims of the water masses. The giants of the sea — whales and dolphins — roam these waters in search of food. The oceans, the "climate breathers", carry heat from the equators to the polar ice caps.

The abyssal depths are dark realms of mystery, an overlooked world. There, in the blackness, strange creatures have thrived, adapted to total darkness and immense pressure. Bioluminescent creatures, giant tube worms,

Fountain of riches

cave-dwelling fish and exotic organisms — these deep-sea survivors show the extreme limits to which nature can push itself.

Ocean trenches, such as the Mariana Trench, present the ultimate challenge. The creatures of these abysses have adapted to harsh temperatures and crushing pressures. Yet life triumphs in these inhospitable surroundings.

Most of the oxygen we breathe comes from the marine environment. From mangroves to coral reefs, marine ecosystems, particularly algae and phytoplankton, convert light into oxygen through photosynthesis, a process unique to the plant kingdom. They purify the atmosphere by filtering carbon dioxide, helping to mitigate the effects of climate change.

For humans, the ocean floor is a treasure chest of natural resources, an invaluable source of wealth. It contains everything from black gold to natural gas, precious minerals and rare earth elements. These colossal resources have a huge impact on industry, energy and the global economy.

Black gold, or oil, found in the depths of the ocean is one of the most coveted resources. Offshore oil is a vital source of fuel that contributes to the growth and development of our economies. It powers our modern societies, fuels our vehicles, heats our homes and drives our industries. However, its extraction and use raises questions about sustainability, marine conservation and the need to find greener energy alternatives, as it is a geological resource that does not renew itself on a human timescale. Ruthlessly exploited, it is destined to disappear.

Natural gas, another bounty of the sea, is used to generate electricity, heat our homes and is increasingly being exploited as a "clean" energy source. Offshore natural gas reserves are crucial to meeting growing energy demand while reducing greenhouse gas emissions. However, its extraction requires advanced technologies and responsible management to minimise environmental risks.

Precious minerals, such as rare metals and diamonds, are found on the seabed. Extracting them is a major technical challenge, but these materials play a vital role in the technology industry, from electronics to renewable energy. Deep-sea mining raises environmental concerns due to its potential impact on fragile marine ecosystems.

The oceans are the arteries of international trade and communication routes that have linked nations since the dawn of human history. Ocean-going vessels have played a vital role in transporting goods, cultures and stories across the waves. They have built links and bridges between distant lands, enabling the exchange of goods, ideas and knowledge between civilisations.

Fountain of riches

The Phoenicians sailed the Mediterranean, the Vikings explored the waters of the North Atlantic and the Silk Road merchants sailed the Indian Ocean. These maritime routes facilitated the trade of spices, silk, precious metals, works of art and countless other goods that enriched cultures and economies around the world.

Over the centuries, shipping evolved from wooden sailing ships to modern steel cargo ships. These advances have greatly accelerated international trade and played a key role in economic globalisation. Today, giant container ships carry billions of tonnes of goods across the oceans, linking continents and nations.

Maritime travel has also driven cultural and human exchange. Sailors, explorers and immigrants crossed the oceans, bringing with them their customs, languages and traditions. Ports and coastal cities became bustling hubs where people from different backgrounds met and formed new communities.

Seafaring has also shaped the history and geopolitics of the world. By crossing the oceans, great maritime powers exerted their influence by establishing colonies, trade routes and empires. Maritime conflicts and alliances have shaped the course of history, from the Hundred Years' War to the Age of Discovery and the two World Wars.

Today, international trade relies heavily on maritime transport, and the importance of the oceans as trade routes continues to grow. With this comes challenges such as plastic pollution, overfishing and climate change.

Coastlines and marine horizons have always held a special attraction for those seeking recreation. They offer treasured retreats, escapes from the daily grind into worlds of infinite promise. Sandy beaches, deep-sea diving, wind-blown sails and vast expanses of ocean are jewels that draw crowds of travellers and contribute to thriving economies in coastal regions.

Pristine sandy beaches, gently lapped by waves and lined with palm trees, are popular destinations for those seeking to relax, recharge and escape the stresses of modern life. These tranquil spaces offer the chance to soak up the sun (just enough for the body to synthesise vitamin D while avoiding sunburn), curl up with a good book or take a soothing walk along the shore. Seaside resorts and coastal communities thrive thanks to these visitors in search of tranquillity and well-being.

But beaches are only one aspect of the ocean's appeal. Adventure seekers can snorkel to the depths to discover dazzling marine life and colourful coral reefs. Scuba diving offers the chance to swim with a variety of creatures, explore mysterious shipwrecks and experience unforgettable moments. Coastal communities benefit from these underwater explorers, drawn by the beauty of the deep.

Fountain of riches

For sailing enthusiasts, the ocean's horizon is the perfect canvas on which to paint their dreams. With sails billowing in the wind, sailors navigate the oceans and discover places inaccessible by land. Regattas, cruises and sea adventures are all occasions to celebrate the splendour of the marine horizon. Marinas and maritime facilities support this passion for sailing.

The vastness of the open sea evokes a sense of freedom and adventure, attracting travellers eager to embark on epic journeys. Oceanic cruises, boat crossings and expeditions are an invitation to appreciate the majesty of the seas.

≈≈≈

Coastlines and sea horizons serve as muses for artists, reflecting nature in its perpetual evolution. The tides, like human emotions, follow their own rhythm, rising and falling in a cosmic dance. This eternal rhythm reminds the human soul that there is an underlying order, even amidst the chaos of the elements.

This majesty is not without danger. While the sea gives life, it can also take it away. Fierce storms, tidal waves and devastating tsunamis are powerful reminders of its power. This duality — between beauty and terror, serenity and chaos — inspires countless stories, poems and works of art that explore the mysteries of creation.

The sea is a canvas on which the reflections of existence are painted. It is a benefactor, a bridge between civilisations, a mirror of nature's moods and a reminder of our place in the world's theatre. Its story endures through the ages. It is a light that shines through the darkness of the past, a compass for the seafarers of the future. The treasures of the oceans are a metaphor for the riches of the human soul — generous, profound, mysterious and infinite.

≈≈≈

10. An inexhaustible source of inspiration

Nothing is more inspiring than the sight of the waves.
George Sand

The sea has always inspired and comforted the human soul. It has the unique ability to evoke feelings of serenity and wonder, becoming a faithful companion to those seeking peace in the midst of life's trials. Over the centuries, its aura of inspiration has guided the pens of artists, poets and philosophers, the brushes of landscape painters and the tools of sculptors.

Its inherent gentleness and the steady rhythm of its tides evoke a soothing melody that cradles our primal instincts. The soft murmur of the waves lapping gently against the shore can cast a hypnotic spell, lulling souls into a state of bliss. For many, the sea is an escape from the hustle and bustle of everyday life, a sanctuary where peace and solitude can be found. Simply sitting on the sand and gazing at the horizon becomes a meditative experience, allowing the mind to relax and find inner peace.

Beyond its calming effect, the sea's ever-changing moods and vast palette of colours are an endless source of creativity. From breathtaking sunsets that paint the sky with fiery hues to tranquil moonlit nights when the ocean shimmers like a starry sky, the splendour of the sea has been the subject of countless works of art, poetry and harmonious melodies.

Its vastness and unpredictability mirror the twists and turns of the human journey, with its ups and downs, challenges and triumphs. The idea that life, like the sea, has its cycles and rhythms, even in the face of adversity, offers a sense of comfort — the certainty that there is a natural course that brings renewal and hope.

In the embrace of its soothing calm, the sea is a sanctuary for those seeking peace and tranquillity, a canvas for creative expression and a symbolic reflection of life's journey, for its timeless beauty and natural rhythms are a manifestation of the truth born of the earth's vital forces.

≈≈≈

When the wandering soul stands before the vast azure expanse of the ocean, a mystical power envelops them, a soothing melody for the restless mind. It is a place where time itself slows

10. An inexhaustible source of inspiration

down, where the turmoil of daily life dissolves into gentle waves that cradle the troubled soul.

The endless rhythm of the sea reveals the secrets of tranquillity. The whispering ebb and flow of the waves becomes a therapeutic melody, a lullaby that soothes anxiety and heals inner turmoil. It is an elixir of calm that soothes the troubled mind and wipes away the scars of stress and anxiety.

The salty breath of the ocean carries a pure essence that cleanses the mind like a stream of fresh spring water. It banishes dark thoughts and unnecessary worries. A serene kiss from the cosmos, the sea breeze brings a breath of fresh air that invigorates and awakens the dormant vitality of the heart.

The sight of the horizon sparks the imagination and invites dreams of unexplored worlds. The mind finds refuge in the contemplation of the endless sea, letting its thoughts drift far into the unfathomable depths. The ceaseless dance of the waves erases the burdens of the past and reveals a space conducive to meditation and reflection.

The sea is a purifier of the soul, a source of peace. Its relentless tides teach patience, much like a guide for the soul seeking harmony. The waves break and recede, whispering a simple truth: even the fiercest storms come to an end.

It is a sanctuary where the weary spirit can renew itself. It offers an escape from the chains of the world, a balm for the wounds of the soul and a promise of healing. It is a melodious song of serenity that resonates through the ages, a reminder of the beauty that lies in simplicity.

≈≈≈

Poets, artists and musicians have often turned to the sea to express the symbolism of passion. In poetry, the sea is perceived as an ardent lover, both beautiful and wild, while the waves crashing against the shore evoke the turmoil of human emotions. Looking out to the horizon, poets see the reflection of their most intimate desires, an eternal quest for the unattainable. The sea, in its many moods, becomes a mirror of the intensity of human emotions.

In literature, the sea is an allegory for passions, a turbulent ocean of emotions that rise and fall like stormy waves. Writers describe the agony of love with the metaphor of rising tides, the joys of discovery with endless horizons, and the tragedy of loss with devastating storms. In *The Tempest*, Shakespeare chose a magical island and a shipwreck to represent the chaos and rebirth of the inner world.

The ocean has served as the backdrop for powerful stories exploring the human struggle against the forces of nature. In *The Toilers of the Sea*, Victor Hugo transports the reader to the world of the Channel Islands, where the

10. An inexhaustible source of inspiration

protagonist, Gilliatt, battles the elements to save a stranded ship. The work highlights human endurance and ingenuity in the face of the vastness and harshness of the sea, making Gilliatt a tragic hero whose devotion to his task reflects his connection to the elements.

Ernest Hemingway's *The Old Man and the Sea* tells the epic story of Santiago, an old Cuban fisherman who fights a giant marlin in the waters of the Gulf Stream. The story explores human dignity in the face of adversity, the endurance of the human spirit in the face of loneliness and the relentless forces of nature. Here, the sea, both a formidable adversary and a source of sustenance, symbolises the eternal cycle of struggle and survival.

In Herman Melville's *Moby Dick*, the sea becomes the setting for an obsessive quest. Captain Ahab's pursuit of the legendary white whale, Moby Dick, embodies the struggle between man and nature and the exploration of the darkest corners of the human soul. The sea, vast and unfathomable, represents the infinite and the unknown, a domain where Ahab's obsession and madness collide with the unyielding power of nature.

In Jules Verne's *Twenty Thousand Leagues Under the Sea*, readers are taken on an extraordinary adventure aboard the submarine *Nautilus*, commanded by the enigmatic Captain Nemo. The novel follows Professor Pierre

Aronnax, his servant Conseil and the harpooner Ned Land, who are captured by Nemo after being sent on an expedition to investigate a supposed sea monster terrorising the oceans.

Instead of finding a monster, they discover the *Nautilus*, an advanced submarine that reveals the hidden wonders of the ocean depths. During their underwater journeys, the characters explore mythical places like Atlantis, meet exotic sea creatures and uncover underwater treasures. Captain Nemo, a mysterious character, is both a scientific genius and a tormented man whose motivations remain unclear.

The novel is a reflection on nature, science and freedom, as well as a critique of the imperial powers of the time. Verne uses the sea as an infinite space of discovery, a place where human laws no longer apply, but where the fascination of the unknown and scientific exploration lead to extraordinary achievements.

Although different in their approach, these works use the sea as a mirror of the passions, struggles and challenges of human existence. They show how the sea, with its mysteries and dangers, continues to inspire narratives about the human condition, illustrating the eternal struggle between man and forces beyond his control.

10. An inexhaustible source of inspiration

Painters and artists have captured the passion of the sea on canvas. The changing colours of the ocean, from dark blue to the golden hues of dusk, embody the different moods of passion. The crashing waves evoke the fire of romantic fervour, while distant horizons symbolise hope and dreams. The works of art also explore the duality of the sea, revealing a rich palette of harmonious contradictions. Paintings depict the tranquillity of a calm sea or the unleashed power of the waves, translating this duality into a visual meditation on life, where seascapes, capturing moments of introspective calm or bursts of passionate expression, become reflections on the human condition.

Several famous paintings use the sea as a backdrop, each offering a unique perspective that contributes to their fame. *The Great Wave off Kanagawa* (1831) by Katsushika Hokusai is one of the most iconic works of Japanese art. It depicts a massive wave threatening to engulf fishing boats, with Mount Fuji in the background contrasting the calm and violence of the sea. The work has become a symbol of Japanese art because of its innovative use of perspective, the dramatic power of the wave and its mastery of woodblock printmaking. It has also been a major influence on many Western artists.

The Birth of Venus (1486) by Sandro Botticelli is another iconic work. This Renaissance painting depicts Venus, the goddess of love, emerging from the sea on a shell carried by gentle winds. The work is celebrated for its idealised beauty, the grace of its figures and its depiction of classical mythology. Reflecting the aesthetic ideals of the Renaissance, Botticelli captures a balance between the human and the divine.

Impression, Sunrise (1872) by Claude Monet, which depicts a sunrise over the harbour of Le Havre with the orange tones of the sun reflected in the sea and some silhouettes of boats, is famous for giving its name to the Impressionist movement. This work captures the fleeting and instantaneous, key characteristics of Impressionism, through its innovative use of light, colour and the technique of painting outdoors.

The *Raft of the Medusa* (1818-1819) by Théodore Géricault represents the survivors of the wreck of the frigate *Méduse* drifting on a raft in the middle of the ocean. The painting is famous for its brutal realism, emotional intensity and social commentary on the incompetence of the French government at the time. Géricault highlights human tragedy with powerful composition and meticulous technique.

The Tempest (1596) by Giorgione depicts a stormy landscape with a town in the background, a soldier standing on the left and a

10. An inexhaustible source of inspiration

woman nursing a child on the right, all under an ominous sky. Famous for its mystery and atmosphere, *The Tempest* is often studied for its innovative use of landscape and light. The exact meaning of the scene is still debated, which adds to the aura of mystery surrounding the work.

A famous painting by Joseph Mallord William Turner depicts a snowstorm at sea. Titled *Snow Storm: Steam-Boat off a Harbour's Mouth* (1842), it shows a steamship battling a violent storm. The tumultuous waves and tormented sky are rendered with a dramatic, almost abstract intensity, creating a sense of natural chaos and the raw power of the sea. Turner was renowned for his mastery of light, colour and movement, and this work perfectly illustrates his almost pre-Impressionist style of painting. He does not simply paint a scene, but conveys the raw emotion evoked by the storm. According to some stories, Turner even asked to be tied to the mast of a ship during a storm in order to better capture the experience he wanted to depict. The work expresses both the sublime grandeur and terror of nature and evokes the unbridled power of the elements and the fragility of humanity in the face of these forces.

These paintings are famous not only for their artistic technique and composition, but also for their ability to capture emotions and dramatic moments using the sea as a central or background element. They continue to inspire and fascinate with their beauty and cultural significance.

In music, the sea has inspired evocative works. Claude Debussy's symphonies, such as *La Mer*, capture the ebb and flow of the waves. These compositions transport the listener to the edge of the ocean, evoking the calm and therapeutic qualities of the sea. It is said that he completed this work during a stay in the seaside town of Eastbourne, England, where he was drawn to the soothing and rhythmic waves.

Composers have created works such as *La Tempesta di Mare*, inspired by the dramatic power of a stormy sea. These compositions share a common inspiration: the evocation of the wild sea, translated into music with an intensity that reflects the raw power of nature. Antonio Vivaldi, for example, composed a flute concerto called *La Tempesta di Mare*. This concerto is particularly famous and is part of the *Concerti di Parigi*, a series of six concertos. Antonio Salieri, another Italian composer of the Classical period, known for his operas and court music, wrote an overture with the same title, which musically captured the turbulence of the sea. Giuseppe Antonio Brescianello, a Baroque composer, captured the essence of a stormy sea in his Sinfonia of the same name. Johann David Heinichen, a German Baroque composer, also wrote a sinfonia on the theme. Although written

10. An inexhaustible source of inspiration

by different musicians, these works share a common inspiration: the evocation of a stormy sea through music.

Sea shanties, on the other hand, tell stories of love lost at sea, of voyages, adventures and loneliness. With its dreams and hopes, the ocean is an ode to life and passion.

Art, like the sea, knows no geographical or cultural boundaries. Artists from around the world, each with their own unique visual language, seize the essence of the sea in a way that resonates universally. Like the sea, art offers an aesthetic experience that can be both soothing and powerful, a means of emotional exploration through which humanity celebrates the duality of life.

≈≈≈

It is not hard work that makes the world go round. While it is essential for turning ideas into reality and projects into reality, it merely materialises what inspiration has first conceived and created. Inspiration, that creative impulse that sometimes comes unexpectedly, is the true engine of progress. It is the source of great discoveries, timeless works of art and innovations that transform our lives.

This creative impulse manifests itself in all aspects of life. It combines and transcends the emotional and rational dimensions of being. It finds its purest expression in art, which translates emotions, visions and dreams that elude words and logic. The artist draws from this inexhaustible source of inspiration to create works that capture the essence of the human experience, touch the soul and resonate through time. It is this creativity that allows us to see beyond the mundane, to transform the ordinary into the sublime, and to open up new perspectives on the world.

In science, this creative impulse takes on a different form. Inspiration becomes innovation and intuition, leading researchers to new hypotheses and solutions. Creative energy is expressed within a rigorous framework where every idea must be tested, proven and validated. But even in this realm of strict logic, it is the spark of inspiration that ignites the greatest breakthroughs, enabling us to push the boundaries of the known and explore the unknown. It is this synergy between creativity and the scientific method that has led to technological revolutions and discoveries that have changed the course of history.

Whether in art or science, inspiration is the engine that drives humanity forward. It unites heart and mind. It combines the emotional and rational aspects of human nature and creates a balance where beauty and truth coexist. Hard work, though indispensable, is merely the tool that gives shape to these visions. Without the

10. An inexhaustible source of inspiration

initial spark of inspiration, progress would be impossible.

Inspiration enables humanity to cross critical thresholds, to make quantum leaps in the search for truth. It is what opens new horizons, pushes us to see beyond established boundaries and gives us the courage to dream bigger. Where hard work can tire in repetition, inspiration breathes new energy. It is a vision that guides efforts towards extraordinary achievements.

The world moves forward through those moments of clarity and genius when a new idea emerges and disrupts established knowledge and practice. Inspiration is the breath that revives the human spirit. It enables us to see what has never been seen, to imagine what has never been imagined. It is through inspiration that humanity can push the boundaries of the possible and reinvent itself.

The sea, vast and untamed, has long been a source of such inspiration. Its endless horizons and ever-changing nature inspire both wonder and reflection, reminding us that just as the tides continually reshape the shore, so too do we have the capacity to change and grow. The ocean's depths hold secrets yet to be discovered, mirroring the unexplored potential within each of us. Just as sailors of old braved uncharted waters, inspiration can guide us to explore new paths, seek innovative solutions and face the challenges of our time with renewed hope.

HEAVEN'S MIRROR

≈≈≈

11. The power of passion

The sea is a raging mirror.
Arthur Rimbaud

The sea, a mirror of our emotions and our search for meaning, embodies both chaos and serenity, loss and rebirth. It symbolises the passions and eternal changes of life. This is why explorers of the soul continue to turn to the sea for inspiration, to tell the story of humanity and to express our essence, far from appearances, in all its grandeur and diversity.

External forces affect our inner ocean. Relationships, places and events exert pressure on our emotions. The cycles of change guide the seasons of the soul. Like the sea ebbing and flowing, our emotions ebb and flow with the constant currents of our lives. Moments of joy follow waves of sadness, and embracing this natural rhythm brings a salutary calm.

Storms do not break the sea; life's challenges do not break the soul. We are shaped by turbulence, but we persevere, evolve and grow. Beneath the raging waves, the sea retains its inner calm. In the same way, one can access serenity in the midst of emotional turmoil. This is a valuable lesson: at the heart of the storm, inner peace endures.

The ocean has always symbolised the passions and eternal metamorphoses of life. Its turbulent waves echo the agonies of the human soul and express the many facets of emotion.

Passion can easily be compared to the violent waves that crash with unparalleled intensity. The human heart is a whirlwind of emotion, from burning love to consuming rage. The ever-changing tides evoke the constant pulse of passion, the relentless highs and lows of desire and ardour.

≈≈≈

Human passions are powerful forces that guide and shape our lives. They emerge from mysterious depths, a complex mix of biology, psychology and environmental influences. They reflect our most intimate desires, the pulse of our being that drives us to dream, to create, to love and to fight.

But where do these passions come from? They often come from an encounter, an experience, a moment when something in us is awakened, ignited like a flame. Perhaps it is a breathtaking view that leaves us speechless, a melody that resonates with our innermost feelings, a book that opens up a new perspective, or a glance that

11. The power of passion

makes our heart beat a little faster. These moments trigger emotional and chemical reactions within us, a subtle alchemy that nourishes and intensifies our drives.

The moon, with its soft and shifting light, often inspires contemplative dreams and passions. The distant twinkling stars awaken in us a thirst for infinity, an insatiable curiosity about the mysteries of the universe. The wind, an invisible caress, can fan the flames of our desires, reminding us that everything is in motion, that nothing is permanent, that our lives are like leaves carried by the breeze. The earth itself, with its cycles and rhythms, its changing seasons, is a constant reminder of our intimate connection with nature.

Can these passions be explained by the biochemical mechanisms of the brain and heart? Only partially. The brain, with its labyrinths of neurons and cascades of neurotransmitters, controls our emotions, desires and impulses. Love, for example, can be broken down into chemical reactions: dopamine, which brings a sense of pleasure and reward; oxytocin, which strengthens emotional bonds; adrenaline, which makes our heart beat faster when we are excited or frightened. The heart, often symbolised as the seat of emotions, takes part in this biochemical symphony; its rhythm speeds up or slows down in response to our feelings.

But human passions go far beyond a simple chemical equation. They are shaped by our culture, personal experiences, memories and aspirations. They are a meeting point between body and mind, between the material and the immaterial. They are nourished by our imagination, by our ability to dream, to hope, to fear.

While science can explain certain aspects of passions, it cannot capture their full essence. Passions are also movements of the soul, silent calls to what is beautiful, what is great, what gives meaning to our existence. They are the invisible engines that drive us to surpass ourselves, to seek further.

These intense and powerful emotions, which can strongly influence an individual's thoughts, behaviour and morality, represent inner forces that are often irrational and can dominate will and reason. Since ancient times, philosophers have questioned the nature of the passions, their role in human life and their relationship to reason.

For Plato and the Stoics, passions were often seen as disruptive elements, affections of the soul that needed to be controlled in order to achieve wisdom and virtue. They believed that passions could lead the soul in directions contrary to reason, and therefore to destructive or immoral behaviour. The ideal for the Stoics

11. The power of passion

was *apatheia*, a state of emotional detachment in which reason rules over the passions.

Interestingly, the modern word "apathy", which is derived from this Greek concept of *apatheia*, has taken on a very different meaning. Whereas *apatheia* was originally understood as a positive state of inner tranquillity, governed by reason and free from the disturbances of unrestrained emotion, apathy now suggests an energised state of indifference. Rather than representing control over passions, apathy now implies a lack of feeling altogether – a detachment from the emotional and intellectual pursuits of life. In this sense, apathy can be seen as a loss of vitality, where the removal of passion leads not to wisdom but to inertia. Rather than offering a balance between reason and emotion, it often signifies a void in which both have been diminished, leaving the individual not in a Stoic sense of clarity, but in a state of disconnection from the world around them.

Aristotle, on the other hand, believed that passions were not inherently bad, but needed to be tempered by reason. In his Ethics, he proposed the doctrine of the "golden mean", where the passions are balanced so that they are neither excessive nor deficient. Thus, anger is not bad if it is felt for the right reasons and within reasonable limits.

In more recent times, philosophers such as René Descartes looked at passions from a different angle. They saw them as movements of the soul caused by the interaction between body and mind. Descartes, in his treatise *The Passions of the Soul*, sought to understand them in order to better control them, claiming that passions, if well managed, could be beneficial by motivating action and enriching life.

For other thinkers, such as David Hume, passions were central to human motivation and even preceded reason in decision-making. Hume saw passions as the true drivers of human action and famously stated that "Reason is, and ought to be, but the slave of passions".

Beyond the risks of irrational impulses, passions are the heartbeat of our humanity, the notes in the symphony of our existence. They are what make us truly alive, what connect us to the universe, to the moon, to the stars, and to this earth that sustains us. They are the eternal flames that light our way and make every moment a search for meaning and beauty.

≈≈≈

12. The unattainable horizon

> *The horizon is in our eyes, not in reality.*
> Angel Ganivet

An imaginary and mysterious line connects the vastness of the sea with the ethereal expanse of the sky. It is where the colours of the earth intertwine in a visual symphony, constantly changing, yet each retaining its own identity.

Although the horizon seems to be within reach, it remains elusive and resists our efforts to grasp it. It embodies our dreams of the unknown while reminding us of the ephemeral nature of our aspirations. Looking towards this infinite horizon evokes a sense of both hope and melancholy. We are drawn by the promise of new explorations, discoveries and fresh perspectives, while at the same time we are aware of the unbridgeable distance between us and that point of escape.

By its very nature, this horizon invites us to pursue it relentlessly, to push further and further in search of meaning and beauty. Existence is an endless exploration, an eternal quest to find meaning in our journey and to push the boundaries of our understanding. Our desire to explore, our thirst for the unknown and our will to unravel the mysteries of the world are reflected in our fascination with this elusive line.

The horizon reminds us that the beauty of existence lies not in the final destination, but in the journey itself. Every day, every sunrise and sunset, offers a new invitation to contemplation, a chance for renewed wonder. It encourages us to embrace the present moment, to appreciate the richness of life and to celebrate the splendour of the world.

One could spend a lifetime moving towards it without ever reaching it, because it is constantly shifting and evading our gaze. This notion of an unattainable point is similar to the mathematical concept of infinity, a concept that by its very nature cannot be fully grasped. As we contemplate this distant line, the mind is led to contemplate infinity, which in turn leads us by natural association to the spiritual. Infinity is not just a number or an abstract idea, but a bridge to questions about our existence, our place in the universe and the mysteries that elude rational thought.

The harmonious union of sea and sky evokes both the pursuit of our dreams and the journey of our aspirations. Our quest for personal fulfilment often begins with setting distant goals, ambitions we hope to achieve one day. This

12. The unattainable horizon

horizon becomes a metaphor for a future that seems both near and unattainable, drawing us in with the promise of undiscovered adventures and unexplored paths.

When we turn to the horizon at sea, we feel a complex mixture of hope and uncertainty. It is a reminder that realising our dreams is not a straight line, but an adventure full of obstacles and challenges. Like the horizon, our dreams can sometimes seem out of reach. But this only strengthens our resolve to pursue them.

Like the sailors who hoisted their sails to cross unknown seas, we can summon the courage to follow our dreams. No matter where our journeys take us, the true essence lies in the enlightenment we feel with each step. It is our duty to keep the light of our dreams alive, for if we lose sight of them, we find ourselves powerless against destiny. In them we find the essence of life: an adventure that rewards us with moments of discovery, growth and fulfilment.

≈≈≈

HEAVEN'S MIRROR

13. The ocean in a drop of water

Emotions are the only thing that we have.
Patricia Highsmith

On the shore, cradled by the stillness of a sleepy day, the arrival of the gentle waves is so subtle that it almost escapes our notice. Then they retreat and continue their incessant ebb and flow, a cosmic danse directed by universal forces. The grains of sand kissed by the sea are silent witnesses to this endless ritual.

Each wave, each movement of the ocean, draws a unique line on the beach. They stretch and recede with a mystical grace and create patterns that defy repetition. Although they come from the same source, no two are alike, and the curves etched into the sand form lines in an ephemeral work of art.

The sea itself is a tireless artist who constantly invents new designs. This eternal movement reflects the constancy of change in the world. It teaches us that even seemingly humble things, like the waves breaking on the shore, are a source of wonder.

On the shore, waves ending their journey evoke the unbroken pulse of existence, the edge of an eternal rhythm where every particle is a wave, a unique experience in the symphony of life.

Each being enters the world with its own path, its own unique experiences and its own personal mark to leave. As the grains of sand welcome each wave, so the earth welcomes life at its birth. Every step we take and every choice we make leaves an imprint. Although it may seem fleeting, it influences the environment and subtly shapes the way the world is formed. Sometimes invisible to the naked eye, these imprints shape the memories, relationships and stories we weave with others. Just as the sea shapes the coastline over time, our actions, whether small or large, shape the landscape of our lives and those of others. This power of influence, however slight or fleeting, testifies to the importance of every moment, every interaction, for all these moments add up to the essence of our passage on earth.

Life is an infinite series of moments, each fundamental in its own way, however fleeting. Although we are all unique, we are united by the vast ocean of humanity. We are all affected by the tides of life, by the events and experiences that touch us. Just as the waves reflect the sunlight, we reveal our own unique brilliance.

13. The ocean in a drop of water

Each generation follows the last, just as each wave follows the one before it. This reminds us of our responsibility as living beings: to care for the earth that welcomes us and to preserve the beauty and harmony of our world, just as every grain of sand retains the imprint of the waves.

Rumi's phrase "You are not a drop in the ocean, you are the ocean in a drop" offers a unique perspective on individuality and interconnectedness. At first glance, the analogy of a drop in the ocean is a common metaphor for insignificance within a larger whole. However, it encourages us to shift our perception. Instead of seeing ourselves as small, isolated entities, we can recognise that within us we embody the whole of the ocean.

This concept suggests the idea of universality within the individual. Each person is a microcosm of the larger cosmos, containing the same essence and potential found in the vastness of the universe. The drop is not separate from the ocean; it embodies it in its essence.

Acknowledging the limitless aspects of ourselves encourages us to tap into the vast reservoir of wisdom, creativity and spirituality that resides within. We are not mere specks lost in the immensity of existence; we carry within us the depth of the infinite.

Furthermore, if each individual is the ocean in a drop, then the essence that connects us is universal. This suggests that the same life force, consciousness and energy flows through everything and everyone.

In a world that emphasises individuality at the expense of interconnectedness, each person, like a drop, has the potential to influence and impact the ocean of humanity. A vast source of power, this legacy comes with great responsibility, which it is our duty to embrace and respect.

≈≈≈

14. Perpetual motion

The sea never gives in, and it teaches us not to give in ourselves.

Eschyle, Persians

An untamed expanse that stretches to infinity, the sea whispers the secrets of the universe to those who know how to listen to its sacred song. It embodies both serenity and storm, the unfathomable vastness and the impulses of the human soul. Like the waves that break with unprecedented power on the shore before receding in a gentle murmur, our emotions reflect the complexity of existence: they fluctuate, surge and calm.

The stillness of the sea, with its endless blue waters and unfathomable depths, invites us to meditate and contemplate our own being. In these moments of tranquillity we find inner peace, a harmony that resonates with the beating of our hearts. The sea, a mirror of our soul, offers us a space to explore the depths of our mind and reconnect with our purest essence.

But the sea is also the stage for storms, where the forces of nature are unleashed with unrelenting force. These moments of turmoil, when the waves rise like mountains and the wind howls in fury, are a reminder of the burning passions and uncontrollable impulses that can stir our souls. In these tumultuous moments, we are confronted with the raw power of our emotions, the wild force that lies within us.

Yet it is in this union of calm and storm, of peace and passion, that the true beauty of the sea lies. It teaches us that life is a constant flow of serene moments and upheavals, that harmony lies in the balance of contrasts. Like the tides, our emotions ebb and flow, shaping our being and reminding us that we are both fragile and powerful, peaceful and passionate.

The sea, with its immensity beyond human comprehension, is a metaphor for our inner journey. It invites us to embrace every facet of our existence, to find beauty in moments of calm and strength in moments of storm. As we gaze upon its infinite waters, we discover reflections of our soul; in this echo between the outer world and our inner selves, we find the truth of our humanity.

At once eternal and ever-changing, the sea becomes nature's greatest poem, an ode to the richness of human existence. It reminds us that in the union of opposites, in the marriage of

14. Perpetual motion

silence and power, lies the true essence of life. Losing ourselves in its vastness, we find ourselves, and in this shifting reflection we understand that we are both united and free, like the waves that dance at the whim of the wind, that melt into the sand and return to the vastness that gave them life.

≈≈≈

HEAVEN'S MIRROR

14. Perpetual motion

"Reflecting on the natural world" series

"Reflecting on the natural world" is a literary odyssey through the elements. Each page becomes a vivid tableau in which land, sea, mountain and desert are mirrors of the human soul. These reflections take us on an introspective journey through the majestic landscapes of the earth. The untamed beauty in every corner of nature is a hymn to the impermanence of existence, its constant transformation to preserve the source of wonder, the precarious harmony of the cosmos.

The countryside offers a bucolic setting where there is harmony between man and the land. Through its artistic creations it unfolds reflections of simplicity, gentleness, but also the strength necessary for the soul to flourish. The green fields, the gentle hills, the mottled woods and the dense magical forests become poetic metaphors: they illustrate the power of emotions and the eternal growth of life.

The sea, a vast expanse of passions, echoes in an epic symphony. The turbulent waves, the salty breeze, every marine detail becomes an allegory of the complexity of emotions. From the wild beauty of storms to moments of serenity when the sea seems to merge with the sky, the pages reveal a landscape in constant metamorphosis, reminding us of the variety of emotions that mark our ephemeral existence.

The mountain stands majestically tall, a symbol of grandeur and endurance. Between the snow-capped peaks and verdant valleys, every step in this vertical realm becomes a metaphor for the soul's journey to discover its own inner strength. The craggy ridges and endless horizons reveal the beauty of resilience, quiet contemplation and serene acceptance of our place in the universe.

The desert, in its seraphic silence, unfolds spaces where stark simplicity becomes a lesson in truth. The unchanging dunes and vast arid expanses invite us to meditate on the ephemeral passage of existence. In this infinite expanse, nature itself becomes a symbol of our search for meaning. It reminds us that every grain of sand carries the history of time that has passed.

Beyond descriptions of landscapes and what they evoke in us, "Reflecting on the natural world" becomes a poetic odyssey that explores the soul through the natural elements. The series is the brainchild of a scientist who dares to venture into art and philosophy, attempting a synthesis between fields that have each followed their own divergent paths after the works of the polymaths of ancient times. It aims to shed light on our understanding of the world, opening the way to contemplation of our ephemeral existence and our place in the infinity of the universe.

≈≈≈

The author

Charles Milton-Scott, the literary pseudonym of a multi-talented scholar, is the embodiment of a personality that crosses disciplinary boundaries. A distinguished professor of medical sciences, he has also devoted his life to the exploration of various fields of knowledge, including mathematics, physics and the philosophies of ancient Greece.

His professional journey extends far beyond academic boundaries. Charles Milton-Scott has had an international career, holding positions such as Head of Department, Dean and Vice-President, actively contributing to the development and excellence of the higher education institutions in which he has worked. His insatiable thirst for knowledge led him in the 1980s to an exciting adventure in the heart of Silicon Valley, California, where he brought his expertise to a leading biotechnology company.

Charles Milton-Scott's contribution to scientific research is considerable: he has published nearly 500 papers in prestigious journals and written about ten books in his field - the genetic and molecular basis of complex human disease. He has also written dozens of articles for newspapers and magazines and about thirty books for the general public. By making complex concepts accessible to a wide range of readers, he contributes to the dissemination of scientific knowledge beyond the university.

He is also appreciated for his sense of humour, which he enjoys sharing. His sharp wit and ability to create a pleasant learning environment have earned him the respect of colleagues and students alike. Aware that each individual has specific needs, he adapts by using different teaching methods. He is able to make connections between seemingly disparate disciplines, encouraging broader and cross-disciplinary thinking.

His multidisciplinary profile and ability to combine scientific rigour, philosophical exploration and a playful approach make him a unique personality.

≈≈≈

Books by the author
(Published by Amazon KDP)

- The pillars of arrogance, *Collection of short stories* (2022)
- The rhinoceros and the dove: An integrated perspective on the matter and mind conundrum, *Essay* (2024)
- The heart of the beholder: Thought on beauty, *Essay* (2024)
- A graceful balance: Symmetry, asymmetry and beauty, *Essay* (2024)

In the series *The echo of vanities*

- Volume 1: Mankind in the connected age, *Essay* (2024)
- Volume 2: Realities and dangers of the connected society, *Essay* (2024)

In the series *Reflections on the natural world*

- The colour of tulips (Contemplation of the countryside), *Essay* (2024)

- Whispers of the rustling leaves (Secrets and wisdom of the forest), *Essay* (2024)
- Mirror of heaven (The sea, serenity and passionate power), *Essay* (2024)
- The radiance of the peaks (The mountains, guardians of the Creation), *Essay* (2024)
- The silence of the wind (The desert, spiritual echo of the sands and the stars), *Essay* (2024)

Translation

Akira Fujimoto, *The way of the dragonfly: a path to victory and personal fulfillment*, translated from the Japanese, *Personal development* (2024)

www.ingramcontent.com/pod-product-compliance
Lightning Source LLC
Chambersburg PA
CBHW070148230526
45471CB00002B/575